AMAZON STICK 4K MAX USER GUIDE FOR BEGINNERS 2023 EDITION

A Simple Guide for effectively using the Fire Stick 4k Max Device with Alexa, plus hacks for troubleshooting, and fantastic tips and tricks.

MICHEAL TIGNER

TABLE OF CONTENTS

Introduction

Welcome to the ultimate user guide for the Amazon Fire TV Stick 4K Max! This powerful streaming device has been designed to take your entertainment experience to the next level, with its stunning 4K Ultra HD resolution and support for high dynamic range (HDR) content. With the Fire TV Stick 4K Max, you have access to thousands of movies, TV shows, and live events from some of the most popular streaming services, including Amazon Prime Video, Netflix, Hulu, and Disney+.

In this guide, we will take you through the process of setting up and using your Fire TV Stick 4K Max, from connecting to your WiFi network and customizing your settings, to streaming your favorite content and playing games. With the help of our detailed instructions and illustrations, you'll be able to navigate the device effortlessly in no time.

We'll also show you how to personalize your Fire TV Stick 4K Max with your favorite channels and apps, as well as how to use the Alexa voice control feature to control your Fire TV Stick 4K Max and other compatible smart home devices with your voice.

Furthermore, we will also guide you on how to use the Watchlist feature to keep track of the content you want to watch, how to use the Show Me mode to control the content your kids can access and how to use the Parental Control feature to restrict the content.

Whether you're a first-time user or an experienced streamer, this guide will help you unlock the full potential of your Fire TV Stick 4K Max and make the most of your streaming experience. So, sit back, relax nd let's dive into the world of streaming with the Fire TV Stick 4K Max!

Chapter 1: What is the Amazon Fire TV Stick 4K Max?

The Amazon Fire TV Stick 4K Max is a streaming device that allow users to access a wide variety of content, including movies, T shows, music, and live sports. It is designed to be plugged into a TV HDMI port and connected to the internet via Wi-Fi or Ethernet.

One of the key features of the Fire TV Stick 4K Max is its ability stream content in 4K resolution. This means that users can enj ultra-high-definition video with a resolution of up to 3840 x 216 pixels, providing a more vivid and detailed viewing experience.

The Fire TV Stick 4K Max also comes with Alexa Voice Remot which allows users to control their TV, find content, and even contr other smart home devices using voice commands. It also allows use to access over 500,000 movies and TV episodes from popul streaming services such as Netflix, Prime Video, Disney+, and more

Additionally, the Fire TV Stick 4K Max supports screen mirrorin which allows users to mirror their smartphone, tablet, or PC screen their TV. This feature is useful for gaming, presenting, and shari content with others.

Key Features

Amazon Fire TV Stick 4K Max offers a wide range of features to enhance the user's streaming experience. These key features include:

4K Streaming: The Fire TV Stick 4K Max supports 4K resolution, which means that users can enjoy ultra-high-definition video with a resolution of up to 3840 x 2160 pixels. This provides a more vivid and detailed viewing experience, especially for movies and TV shows that are available in 4K.

HDR and HDR10+ support: The Fire TV Stick 4K Max supports HDR and HDR10+, which means that it can display a wider range of colors and brightness. This results in a more vivid and detailed viewing experience, especially for movies and TV shows that are available in HDR.

Alexa Voice Remote: The Fire TV Stick 4K Max comes with an Alexa Voice Remote, which allows users to control their TV, find content, and even control other smart home devices using voice commands.

Wide Variety of Content: The Fire TV Stick 4K Max provides access to over 500,000 movies and TV episodes from popular streaming services such as Netflix, Prime Video, Disney+, and more.

Screen Mirroring: The Fire TV Stick 4K Max supports screen mirroring, which allows users to mirror their smartphone, tablet, or PC screen on their TV. This feature is useful for gaming, presenting, and sharing content with others.

Personalized Experience: The Fire TV Stick 4K Max allows users to customize their home screen, set parental controls, and manage their subscriptions, resulting in a personalized experience.

Easy to Set Up: The Fire TV Stick 4K Max is easy to set up and comes with a quick start guide in the box, making it simple for users to start streaming their favorite content.

Fast Performance: The Fire TV Stick 4K Max has a powerful quad-core processor and 1GB of RAM which provides fast performance while streaming and navigating through the menu.

Ethernet Connectivity: The Fire TV Stick 4K Max has an optional Ethernet adapter which allows users to connect to the internet via Ethernet cable for a more stable and fast internet connection.

Overall, the Amazon Fire TV Stick 4K Max is a powerful and versatile streaming device that offers a high-quality viewing experience and a wide range of features for users to enjoy. Whether you're streaming movies, TV shows, music, or live sports, the Fire TV Stick 4K Max makes it easy to find and watch your favorite content.

Box Contents

The box contents of the Amazon Fire TV Stick 4K Max include the following items:

- Fire TV Stick 4K Max: This is the main device that is designed to be plugged into a TV's HDMI port and connected to the internet via Wi-Fi or Ethernet.

- Alexa Voice Remote: This is the remote control that comes with the Fire TV Stick 4K Max. It allows users to control their TV, find content, and even control other smart home devices using voice commands. The remote also features volume and power buttons, so you can control your TV as well.

- HDMI Extender: This is a small device that is included in the box to help you plug the Fire TV Stick 4K Max into your TV, even if it's in a tight space.

- USB cable and Power adapter: These are included in the box to provide power to the Fire TV Stick 4K Max.

- Quick Start Guide: This is a guide that provides step-by-step instructions on how to set up and start using your Fire TV Stick 4K Max.

- Optional: Ethernet Adapter: An adapter that allows users to connect to the internet via Ethernet cable for a more stable and fast internet connection, this adapter is included in some of the packages.

All of these items are included in the box and are necessary for setting up and using the Fire TV Stick 4K Max. With the included quick start guide, users can easily set up their Fire TV Stick 4K Max and start streaming their favorite content in no time.

bout the Fire TV Settings

he Fire TV stick settings allow you to customize and configure arious aspects of your device, such as network settings, display ettings, and accessibility settings. Here are some of the settings that ou can adjust on your Fire TV stick:

- **Display & Sound**: Allows you to adjust the display resolution, refresh rate, and audio output settings.

- **Network**: Allows you to connect to a Wi-Fi network, view network details, and configure network settings.

- **Alexa**: Allows you to enable or disable Alexa on your Fire TV stick, and also allows you to link it to your Amazon Echo device.

- **Parental controls**: Allows you to set up a PIN and content filters to restrict access to certain content and features on your Fire TV stick.

- **Device**: Allows you to view device information, update software, and manage device options.

- **My Fire TV**: Allows you to customize the apps and channels that appear on your home screen, and manage your apps, games, and channels.

- **Accessibility**: Allows you to enable and configure accessibili features, such as subtitles, closed captions, and voice guidance.

- **Time**: Allows you to set the time zone and turn on or off th automatic time updates.

- **Profiles & Family Library**: Allows you to create and manag multiple profiles on your Fire TV stick and share your Amazo Prime Video content with family members.

- **System**: Allows you to view system information, manag developer options, and factory reset your Fire TV stick.

Note: The settings available on your Fire TV stick may va depending on the specific model and software version of your devic Additionally, you can access the settings menu by selecting th settings option on the top of the home screen.

Chapter 2: Setting Up Your Fire TV Stick 4K Max

Setting up your Amazon Fire TV Stick 4K Max is a relatively simple process that can be completed in just a few steps. Here's a general overview of what you'll need to do:

- Connect the Fire TV Stick 4K Max to your TV: Insert the Fire TV Stick 4K Max into your TV's HDMI port and plug the USB cable into the power adapter. Then, plug the power adapter into an electrical outlet.

- Connect to Wi-Fi: On your TV, navigate to the Fire TV Stick 4K Max home screen and select the "Settings" option. From there, select "Network" and then "Wi-Fi" to connect to your home Wi-Fi network. If you have an Ethernet adapter you can use that instead.

- Register your Fire TV Stick 4K Max: Once you're connected to the internet, the Fire TV Stick 4K Max will prompt you to register the device. You'll need to enter your Amazon account information (or create a new account if you don't already have one) and agree to the terms of service.

- Update the device: After registration, the Fire TV Stick 4K Max will check for updates and automatically download and install any available updates.

- Personalize your experience: Once the setup process is complete, you can customize your home screen, set parental controls, and manage your subscriptions.

- Start streaming: After completing the setup process, you can start streaming your favorite content. You can access a wide variety of content such as movies, TV shows, and music from popular streaming services such as Netflix, Prime Video, Disney+, and more.

Overall, the setup process for the Fire TV Stick 4K Max is straightforward and easy to follow. With the included quick start guide, users should have no trouble getting their device up and running in no time.

Connecting to Your TV

Connecting the Amazon Fire TV Stick 4K Max to your TV is one of the first steps in the setup process and it is a simple process. Here are the detailed steps for connecting the Fire TV Stick 4K Max to your TV:

- Locate the HDMI port on your TV: The HDMI port is typically located on the back or side of your TV, and it's where you'll connect the Fire TV Stick 4K Max. If you have difficulty finding the HDMI port on your TV, consult your TV's manual for more information.

- Insert the Fire TV Stick 4K Max into the HDMI port: Carefully insert the Fire TV Stick 4K Max into the HDMI port of your TV. Make sure the device is securely plugged in and the HDMI extender is plugged in, if needed.

- Turn on your TV: Make sure your TV is on and set to the correct HDMI input. The input can be changed using the TV remote.

- Wait for the Fire TV Stick 4K Max home screen to appear: Once the Fire TV Stick 4K Max is properly connected to your TV, the device's home screen should appear on your TV. This can take a few seconds, so be patient.

- Connect to Wi-Fi: On your TV, navigate to the Fire TV Stick 4K Max home screen and select the "Settings" option. From there, select "Network" and then "Wi-Fi" to connect to your home Wi-Fi network. If you have an Ethernet adapter you can use that instead.

By following these steps, you should be able to successfully connect the Fire TV Stick 4K Max to your TV. Once the device is connected, you'll be able to start streaming your favorite content in no time.

onnecting to Wi-Fi

onnecting the Amazon Fire TV Stick 4K Max to your home Wi-Fi etwork is an important step in the setup process as it allows the evice to access the internet and stream content. Here are the detailed eps for connecting the Fire TV Stick 4K Max to your Wi-Fi etwork:

- On your TV, navigate to the Fire TV Stick 4K Max home screen and select the "Settings" option.

- From the settings menu, select "Network."

- Select "Wi-Fi" from the network options.

- Select your Wi-Fi network from the list of available networks. If your network is not listed, you may need to manually enter the network name and password.

- Once you are connected to your Wi-Fi network, your Fire TV Stick 4K Max will automatically check for updates and download and install any available updates.

- Once the updates are installed, you can start streaming your favorite content.

- Alternatively, you can use an Ethernet adapter (if you have on to connect to your router via an Ethernet cable. This will provi a more stable and faster internet connection.

It's important to note that the Fire TV Stick 4K Max requires minimum internet speed of 7 Mbps to stream content in 4K, while speed of 15 Mbps or more is recommended for the best streamir experience. If you experience buffering or slow performance, yo may need to check your internet speed or contact your internet servi provider for assistance.

By following these steps, you should be able to successfully conne the Fire TV Stick 4K Max to your Wi-Fi network. Once the device connected, you'll be able to start streaming your favorite content in r time. If you have any trouble connecting to your Wi-Fi network, yo can refer to the quick start guide that comes with the device or conta Amazon customer service for assistance.

How to connect your fire tv stick to a public network

To connect your Fire TV stick to a public network, such as a hotel or public Wi-Fi, you will need to follow these steps:

- Turn on your Fire TV stick and navigate to the home screen.
- Select the "Settings" option located on the top of the screen.
- Select "Network"
- Select "Wi-Fi"
- Select the public network you want to connect to from the list of available networks.
- When asked, provide the network password.
- Select "Connect"
- Once connected, your Fire TV stick will automatically check for updates, if there is any update it will prompt you to download and install it.

Note: Some public networks may require additional authentication or a web login before you can connect to them. Additionally, Public networks may have lower security and may be vulnerable to hacking, it is recommended to avoid sensitive activities like online banking or shopping while connected to a public network.

Connecting the Amazon Fire Stick to WIFI without using the remote

If you don't have access to the remote for your Amazon Fire TV stick, you can still connect it to your Wi-Fi network using the following steps:

- Download the Amazon Fire TV Remote App from the App Store or Google Play Store on your smartphone or tablet.
- Open the app and sign in with the same Amazon account you used to set up your Fire TV stick.
- On your Fire TV stick, go to the settings menu by pressing the home button on the remote three times quickly.
- Select "Device" and then "About."
- Select "Network" and then "Wi-Fi."
- On the app, select the "Device" button at the bottom of the screen.

- On the app, select the name of your Fire TV stick from the list of devices.

- On the app, select "Wi-Fi" and then select your desired Wi-Fi network from the list.

- On the app, enter the password for your Wi-Fi network.

- On the Fire TV stick, select "Connect."

- Once the Fire TV stick is connected to your Wi-Fi network, the app will display a message saying "Connected."

Note: Some steps may vary depending on the version of the app and the version of the Fire TV stick, and it's important to have your device and your firestick on the same network.

Chapter 3: Registering Your Fire TV Stick 4K Max

Registering your Amazon Fire TV Stick 4K Max is an important step in the setup process as it allows you to access your Amazon account and start streaming content. Here are the detailed steps for registering your Fire TV Stick 4K Max:

- Connect to Wi-Fi: Make sure your Fire TV Stick 4K Max is connected to your home Wi-Fi network before proceeding with registration.

- Select "Sign in" or "Register" on the screen: Once the device is connected to the internet, a screen will appear prompting you to sign in or register.

- Sign in with your Amazon account: If you already have an Amazon account, enter your email address and password to sign in.

- Create a new Amazon account: If you don't have an Amazon account, you can create a new one by following the on-screen instructions. You'll need to provide your name, email address, and a password.

- Accept the terms of service: Once you're signed in or registered, you'll need to accept the terms of service.

- Personalize your experience: After the registration process is complete, you'll be prompted to personalize your experience. You can customize your home screen, set parental controls, and manage your subscriptions.

y following these steps, you should be able to successfully register our Fire TV Stick 4K Max. Once the device is registered, you'll be ble to start streaming your favorite content from popular streaming rvices such as Netflix, Prime Video, Disney+, and more.

's worth mentioning that, if you have any trouble with registration, u can refer to the quick start guide that comes with the device or ntact Amazon customer service for assistance. Also, if you have ultiple Amazon accounts, make sure you use the right one to avoid ny issues with streaming and purchasing.

How to configure your fire stick during setup

During the setup process of a Fire TV stick, there are sever configuration options that you can adjust to customize yo experience. Here are some of the configuration options that you ma be prompted to adjust during the setup process:

- Language: Allows you to select the preferred language for yo Fire TV stick's interface.

- Wi-Fi: Allows you to connect your Fire TV stick to your hon network to access the internet and streaming content.

- Registration: Allows you to register your Fire TV stick wi your Amazon account, which is required to access certa features, such as streaming content from Amazon Prime Vid and purchasing apps from the Amazon Appstore.

- Time zone: Allows you to set the correct time zone for your Fi TV stick.

- Display & Sound: Allows you to adjust the display settings such as the resolution, and audio settings such as the audio output.

- Parental controls: Allows you to set up a PIN and content filters to restrict access to certain content and features on your Fire TV stick.

- Alexa: Allows you to enable or disable Alexa on your Fire TV stick, and also allows you to link it to your Amazon Echo device.

Note: The setup process may vary depending on the specific model and software version of your Fire TV stick. Additionally, you can adjust these settings later by going to the "Settings" menu on your Fire TV.

How to configure the language and time

Here are the steps to configure the language and time on your Fire TV stick:

- Turn on your Fire TV stick and navigate to the home screen.
- Select the "Settings" option located on the top of the screen.
- Select "Device"
- Select "Device Options"
- Select "Language" and select your desired language from the list.
- Select "Time Zone" and select your time zone from the list
- Select "Save"

Once you have made your selections, your Fire TV stick will be set to the language and time you have chosen.

Note: Some devices or models may have variations on the menu or location of the settings, but the general process of selecting Language and Time Zone should be similar. Additionally, you can also use the Alexa voice command to change the language, for example, "Alexa, change the language to Spanish" or change the time, for example, "Alexa, set the time zone to Pacific Time".

How to change the name of your fire tv stick

You can change the name of your Fire TV stick to make it more identifiable on your network. Here are the steps to change the name of your Fire TV stick:

- Turn on your Fire TV stick and navigate to the home screen.
- Select the "Settings" option located on the top of the screen.
- Select "Device"
- Select "About"
- Select "Device name"
- Type in the new name you want to give to your Fire TV stick.
- Select "Save"

Alternatively, you can also change the name of your Fire TV stick through the Amazon Alexa app:

- Open the Amazon Alexa app on your smartphone or tablet
- Tap on the menu icon and select "Settings"
- Select "Device Settings"
- Select your Fire TV stick

- Select "Device Info"
- Select "Device Name"
- Type in the new name you want to give to your Fire TV stick.
- Select "Save"

Note: Changing the name of your Fire TV stick will not affect its functionality or performance. Additionally, you can also use the name to refer to your Fire TV stick when giving voice commands to Alexa-enabled devices.

How to change your location on a fire stick

The location on your Fire TV stick is used to determine the availability of certain apps, content and features. Here are the steps to change your location on a Fire TV stick:

- Turn on your Fire TV stick and navigate to the home screen.
- Select the "Settings" option located on the top of the screen.
- Select "Device"
- Select "Device Options"
- Select "Country"
- Select the country you want to change to, then select "Save"
- After you have changed your location, the Fire TV stick will need to be restarted, you will be prompted to do so.
- After restarting, the Fire TV stick will be set to the new location you have chosen.

Note: Changing the location on your Fire TV stick might cause some apps or content to be unavailable, or cause changes to the app store, content library or other services. Additionally, some devices or ıodels may have variations on the menu or location of the settings, ut the general process of selecting the country should be similar.

Chapter 4: Using Your Fire TV Stick 4K Max

he Amazon Fire TV Stick 4K Max is a powerful and versatile reaming device that offers a wide range of features for users to ıjoy. Here are some of the key features and functionality of the Fire V Stick 4K Max:

- **Navigating the Home Screen**: The Fire TV Stick 4K Max's home screen is the starting point for accessing content and features. The home screen displays a list of recommended content and apps, as well as quick access to recently used apps, settings, and other options. You can navigate through the home screen using the remote's directional buttons and select items by pressing the center button.

- **Finding and Watching Content**: The Fire TV Stick 4K Max provides access to over 500,000 movies and TV episodes from popular streaming services such as Netflix, Prime Video,

Disney+, and more. Users can search for content by title, acto or genre, and also browse different categories such as "Recentl Watched," "New Releases," and "Prime Originals." Once yc find the content you want to watch, simply select it and press tl center button on the remote to start streaming.

- **Voice Control with Alexa**: The Fire TV Stick 4K Max come with an Alexa Voice Remote that allows users to control the TV and find content using voice commands. Users can as Alexa to play a specific movie or TV show, search for conte by genre or actor, and even control other smart home devices.

- **Screen Mirroring**: The Fire TV Stick 4K Max supports scree mirroring, which allows users to mirror their smartphone, table or PC screen on their TV. This feature is useful for gamin presenting, and sharing content with others.

- **Personalized Experience**: The Fire TV Stick 4K Max allov users to customize their home screen, set parental controls, ar manage their subscriptions, resulting in a personaliz experience.

Overall, the Fire TV Stick 4K Max is a powerful and versati streaming device that offers a high-quality viewing experience and wide range of features for users to enjoy. Whether you're streamir movies, TV shows, music, or live sports, the Fire TV Stick 4K M makes it easy to find and watch your favorite content.

Using OTG on Fire Tv

OTG, or "On-The-Go," is a feature that allows you to connect USB devices to your Fire TV stick using an adapter. Here are the steps to use OTG on your Fire TV stick:

- Turn on your Fire TV stick and navigate to the home screen.
- Obtain an OTG (On-The-Go) USB adapter that is compatible with your Fire TV stick.
- Connect the adapter to the USB port on your Fire TV stick.
- Connect the USB device you want to use to the adapter.

- Once the device is connected, the Fire TV stick will automatically recognize it and you can access the files or content on the USB device.

Note: The availability of the feature may depend on the specific model of your Fire TV stick. Additionally, not all USB devices are compatible with the Fire TV stick. It's recommended to check the device's documentation or contact the manufacturer for compatibility information. Some devices that can be used with an OTG adapter include USB drives, keyboards, mice, and game controllers.

How to Turn Off Targeted Adverts

Fire TV stick uses your viewing history, search history, and other information to personalize your experience and show you targeted advertisements. Here are the steps to turn off targeted advertisements on your Fire TV stick:

- Turn on your Fire TV stick and navigate to the home screen.
- Select the "Settings" option located on the top of the screen.
- Select "Preferences"
- Select "Privacy Settings"
- Select "Interest-Based Ads"
- Turn off the toggle switch for "Interest-Based Ads."

Alternatively, you can also turn off targeted advertisements by visiting the Amazon Advertising Preferences website and signing in with your Amazon account:

- Open your web browser and go to the Amazon Advertising Preferences website.
- Sign in with your Amazon account.
- Select "Your Advertising Preferences"
- Turn off the toggle switch for "Interest-Based Ads."

Note: Turning off targeted advertisements may result in less relevant ads on your Fire TV stick, but it will also prevent Amazon from collecting your data for ad personalization. Additionally, some apps and channels may still show you ads even if you have turned off targeted advertisements on your Fire TV stick.

How to disable Auto-playing video

Auto-playing videos refer to videos that start playing automatically when you open an app, website, or channel. This feature can be useful in some cases, but it can also be annoying and consume data. Here are the steps to disable auto-playing videos on your Fire TV stick:

- Turn on your Fire TV stick and navigate to the home screen.
- Select the "Settings" option located on the top of the screen.
- Select "Preferences"
- Select "Auto-Play"
- Turn off the toggle switch for "Auto-Play Videos"
- Once you complete these steps, the videos will not play automatically anymore and you will have to manually start them.

Alternatively, you can also disable auto-playing videos for specific apps, channels, or websites by adjusting the settings within the app or website.

Note: Disabling the auto-playing videos will not affect the functionality or performance of your Fire TV stick. Additionally, Some apps or channels may not have an option to turn off auto-playing videos and in that case, you will have to use the general option to disable it.

How to Play Games on Fire Tv

The Fire TV stick allows you to play games on your TV using the included remote or a compatible game controller. Here are the steps to play games on your Fire TV stick:

- Turn on your Fire TV stick and navigate to the home screen.
- Select the "Apps" option located on the top of the screen.
- Scroll down to the "Games" option and select it.
- Browse the available games or use the search bar to find specific games.
- Select the game you want to play and select "Download" or "Buy" to install it.
- Once the game is installed, select "Open" to launch it.

- Follow the on-screen instructions to start playing the game.
- To use a game controller, connect it to your Fire TV stick via Bluetooth or USB.

ote: Some games may require a game controller to play, while :hers can be played using the included remote. Additionally, you can so subscribe to Twitch Prime and get free games, in-game content, ıd a free channel subscription every month.

Chapter 5: Navigating the Home Screen

he Amazon Fire TV Stick 4K Max's home screen is the starting oint for accessing content and features. The home screen displays a st of recommended content and apps, as well as quick access to cently used apps, settings, and other options. Here's a detailed look how to navigate the home screen:

- **The main menu**: The main menu is located at the top of the home screen and includes options such as "Home", "Apps", "Settings", "Movies", "TV", "Music", and "Search". You can access these options by using the remote's directional buttons navigate and press the center button to select the option.

- **The Carousel**: The carousel is located at the top of the hon screen and displays a list of recommended content and app You can scroll through the carousel by using the remote directional buttons and press the center button to select the iten

- **The Row**: The row is located below the carousel and displays list of recently used apps, settings, and other options. You ca scroll through the row by using the remote's directional butto and press the center button to select the item.

- **The App Grid**: The App Grid is located at the bottom of tl home screen and displays all the apps that are installed on yo device. You can scroll through the App Grid by using tl remote's directional buttons, and press the center button to sele the app.

- **Search**: You can quickly search for content on your Fire T Stick 4K Max by using the remote's microphone button ar speaking into it. You can also use the on-screen keyboard enter your search query by highlighting the search icon on tl main menu and selecting it.

Navigating the home screen on the Fire TV Stick 4K Max is intuiti and straightforward. By using the remote's directional buttons, y can easily access the main menu, the carousel, the row, the app gri and search for content on your device.

How to Use the Fire tv Remote App.

The Fire TV Remote app is a free mobile application that allows you to control your Fire TV stick using your smartphone or tablet. The app may be downloaded from the App Store or Google Play Store and works on both iOS and Android devices. Here are the steps to use the Fire TV Remote app:

- Download and install the Fire TV Remote app on your smartphone or tablet.

- Open the app and sign in with your Amazon account.

- Make sure your Fire TV stick and mobile device are connected to the same Wi-Fi network.

- Select your Fire TV stick from the list of available devices.

- Use the app to navigate the Fire TV stick interface, launch apps, and control playback.

The app offers a variety of features that can enhance your Fire TV stick experience, such as:

- Voice search: Use the microphone button on the app to search for content using your voice.

- Keyboard: Use the virtual keyboard to type text on your Fire TV stick.

- Swipe navigation: Swipe left and right to navigate through the Fire TV stick interface.

- Control playback: Use the app to play, pause, fast forward, and rewind videos.

- Volume control: Use the app to control the volume of your Fire TV stick.

Note: The Fire TV Remote app requires an active internet connection to function. Additionally, the app may not be compatible with all Fire TV models, and some features may vary depending on the specific model and software version of your Fire TV stick.

Tips and Tricks to Get the Best out of your Fire Stick

There are many ways to enhance your Fire TV stick experience, here are some tips and tricks to help you get the most out of your device:

- **Use the voice search feature**: Press the microphone button on your remote control and say the name of the show, movie, or app you want to find.

- **Customize the home screen**: Press and hold the home button on your remote control to access the quick actions menu, where you can rearrange and add your favorite apps and channels to the home screen.

- **Use the closed captions**: Go to settings, select "Accessibility," and turn on closed captions to display subtitles for all videos.

- **Use the Picture-in-picture mode**: While watching a video, press the menu button on your remote control to enter the picture-in-picture mode and continue browsing other content.

- **Use the Amazon Echo to control your Fire TV stick**: Connect your Fire TV stick and Echo device to the same WiFi network and use voice commands to control your Fire TV stick.

- **Use the mirroring feature**: Install the screen mirroring app on your smartphone, tablet, or PC and mirror the screen to your Fire TV stick.

- **Use the Parental controls**: Go to settings, select "Parental controls," and set up a PIN to restrict access to certain content and features on your Fire TV stick.

- **Use the FreeTime feature**: Go to settings, select "Parental controls," and set up the FreeTime feature to restrict the content that your children can access.

- **Use the Amazon Prime Video Channels**: Go to the Amazon Prime Video app, and subscribe to additional channels such as HBO, STARZ, and more, as an add-on to your Prime membership.

- **Use the Amazon Kids+**: Go to the Amazon Prime Video app, and subscribe to Amazon Kids+ to access a wide range of kid-friendly content.

ote: Some of the features may vary depending on the specific model ıd software version of your Fire TV stick. Additionally, the tips and icks mentioned here are not exhaustive and there are many other ays to use your Fire TV stick.

anner features on the Fire TV

anner features on the Fire TV include:

- **Personalized recommendations**: Users can see personalized recommendations for movies, TV shows, and other content based on their viewing history and preferences.

- **Voice search**: Users can use their voice to search for content o the Fire TV.

- **Live streaming**: Users can watch live streaming content from variety of providers, including Amazon Prime Video, Netflix, and Hulu.

- **4K Ultra HD**: The Fire TV supports 4K Ultra HD resolution, allowing users to enjoy high-quality video content on compatible TVs.

- **Alexa integration**: The Fire TV is compatible with Alexa, Amazon's virtual assistant, allowing users to control their TV using voice commands.

- **Parental controls**: Users can set parental controls to restrict access to certain content and set time limits for viewing.

- **Multi-language support**: The Fire TV supports multiple languages, making it accessible to a wide range of users.

Chapter 6: Finding and Watching Content

The Amazon Fire TV Stick 4K Max provides access to a wide varie of content, including movies, TV shows, and music from popul streaming services such as Netflix, Prime Video, Disney+, and mor Here's a detailed look at how to find and watch content on the Fire T Stick 4K Max:

- **Finding content**: To find content on the Fire TV Stick 4K Max, you can navigate to the main menu on the home screen and select the "Movies," "TV," or "Music" option. You can also use the search feature by pressing the microphone button on the remote and speaking into it, or by using the on-screen keyboard to enter your search query.

- **Browsing content**: Once you've selected the "Movies," "TV," or "Music" option from the main menu, you can browse different categories such as "Recently Watched," "New Releases," and "Prime Originals." You can also browse by genres, actors, or directors by selecting the "Filter" option.

- **Streaming content**: Once you've found the content you want to watch, simply select it and press the center button on the remote to start streaming. If the content is available in 4K, it will automatically stream in 4K.

- **Resume watching**: If you've previously started watching a TV show or movie and haven't finished it, you can resume watching it from where you left off by selecting the "Resume" button.

- **Parental controls**: If you have set parental controls on your Fire TV Stick 4K Max, you will be prompted for a PIN before streaming certain content.

It's worth mentioning that, to stream content from some of the services, like Netflix, Disney+, you need to have a subscription to that service, and also, to stream content in 4K, you need to have an internet speed of 7 Mbps or higher.

Overall, finding and watching content on the Fire TV Stick 4K Max is easy and intuitive. With a wide variety of content available and the ability to search and browse by category, the Fire TV Stick 4K Max makes it easy to find and watch your favorite content.

Turning On and Off the Subtitle

Fire TV stick allows you to turn on and off subtitles for your streaming content, here are the steps to do so:

- Turn on your Fire TV stick and navigate to the home screen.
- Start playing a video or movie.
- Press the menu button on your remote (it's the button with three lines)

- Select "Subtitles"
- Choose the subtitle language you want to turn on, or select "Off" to turn off subtitles.
- Press the "Play" button to resume playback with the new subtitle settings.

Alternatively, you can also change the subtitle settings in the settings menu:

- Turn on your Fire TV stick and navigate to the home screen
- Select the "Settings" option located on the top of the screen.
- Select "Accessibility"
- Select "Subtitles"
- Turn on or off the subtitles.
- Choose the subtitle language you want to turn on or select "Off" to turn off subtitles.

Note: The availability of subtitles may depend on the specific content, and not all shows and movies will have subtitles available. Additionally, the process of turning on and off subtitles may vary depending on the content provider and the app you are using.

Finding TV Shows and Movies

There are several ways to find TV shows and movies on a Fire TV stick:

- **Home screen**: On the Fire TV home screen, there are sections for "Featured" and "Popular" TV shows and movies. You can browse these sections to find content that is currently popular or recommended by Amazon.

- **Search**: Press the microphone button on your Alexa Voice Remote and say "search for (TV show or movie title)" or "find (TV show or movie title)" to quickly find a specific title. You can also search for specific genres, for example, "search for action movies"

- **Applications**: You can access the apps installed on your Fire TV, such as Netflix, Amazon Prime Video, and Hulu, to find TV shows and movies available on those platforms.

- **Your Library**: If you have subscribed to a streaming service that allows you to download content, you can find the content you have downloaded in "Your Library" section.

- **Watchlist**: If you have added a movie or TV show to your watchlist, you can find them in the "Your Library" -> "Watchlist" section.

Note: Some apps might have a different layout or menu, you can check the app's menu or settings for more options. Additionally, you can use the "Show Me" feature to find specific types of content, like "Show me romantic comedies" or "Show me live sports".

How to use the Show Me Mode

The "Show Me" mode is a feature on the Fire TV that allows you to quickly access a specific type of content, such as movies, TV shows, or live TV, using voice commands. Here's how to use it:

- Turn on your Fire TV and navigate to the home screen.

- Press and hold the microphone button on your Alexa Voice Remote and say "Show me [content type]", for example, "Show me movies" or "Show me live TV"

- The Fire TV will display a selection of movies or live TV options based on your request.

- You can also filter your request, for example, "Show me romantic comedies" or "Show me live sports"

- You can also use the Show Me mode to access specific apps or channels, for example, "Show me Netflix" or "Show me ESPN"

- Once you've found the content you're looking for, you can use voice commands to control playback and volume, such as "Play" or "Pause" and "Volume up/down"

ote: The Show Me mode feature may vary depending on your cation and the services you have access to, but it can be a handy ay to quickly access the content you want.

low to Find a Workout

ne Fire TV stick allows you to access a variety of workout apps and deos that can help you stay active and healthy. Here are some steps get a workout using your Fire TV stick:

- Turn on your Fire TV stick and navigate to the home screen.

- Select the "Apps" option located on the top of the screen.

- Scroll down to the "Sports & Fitness" category and select it.

- Browse the available workout apps such as "Nike Trainin Club," "PEAR Personal Coach," "Daily Burn," or "Fitbit Coach and select it.

- Select "Download" or "Buy" to install the app.

- Once the app is installed, select "Open" to launch it.

- Follow the on-screen instructions to create an account, set yo fitness goals, and start your workout.

- To use a game controller, connect it to your Fire TV stick v Bluetooth or USB.

Alternatively, you can also search for workout videos on streamin services like Amazon Prime Video, Netflix, or YouTube by searchin for "workouts", "fitness", "yoga", "cardio", etc.

Note: Some apps may require a subscription or a one-time purchas but most of them offer a free version with a limited number workouts. Additionally, it's recommended to consult a physici before starting any new exercise program.

Using the Watchlist

The "Watchlist" feature on a Fire TV allows you to keep track of movies and TV shows that you want to watch in the future. Here's how to use it:

- Turn on your Fire TV and navigate to the home screen.
- Find the movie or TV show that you want to add to your watchlist and select it.

- On the detail page of the movie or TV show, select the "Add to Watchlist" option.
- The item will now be added to your Watchlist, which you can access by going to the "Your Library" section on the main menu.
- To remove an item from your Watchlist, go to the "Your Library" section, select "Watchlist", find the item you want to remove, and select "Remove from Watchlist."
- You can also access your Watchlist on other devices like your phone, tablet, or computer by logging into your Amazon account on the Amazon Prime Video app or website.

Note: To use the Watchlist feature, you'll need an Amazon Prime account and the Amazon Prime Video app installed on your Fire TV. Also, if you have other profiles in your FireTV, the Watchlist is personalized to each profile, so each profile will have its own Watchlist.

How to Turn off Autoplay for Trailer

The "Autoplay" feature on a Fire TV automatically plays trailers for related movies and TV shows after the one you are currently watching. If you prefer not to have trailers automatically play, you can turn off this feature. Here's how:

- Turn on your Fire TV and navigate to the home screen.

- Select the "Settings" option located on the top of the screen.
- Select "Preferences"
- Select "Playback"
- Toggle off "Autoplay trailers and more."

This will prevent trailers from automatically playing after the content you are currently watching, and you will have to manually select the "Play" button to watch a trailer.

Note: This setting will also stop the Autoplay of "related videos" feature, which suggests other videos you may like to watch. Additionally, this setting will be applied to all the profiles in your FireTV, so if you have multiple profiles, you will have to turn it off for each one individually.

Chapter 7: Voice Control with Alexa

The Amazon Fire TV Stick 4K Max comes with an Alexa Voice Remote that allows users to control their TV and find content using voice commands. Here's a detailed look at how to use voice control with Alexa on the Fire TV Stick 4K Max:

- Press and hold the microphone button on the remote: To activate Alexa, press and hold the microphone button on the remote. A blue light will appear on the remote to indicate that Alexa is listening.

- Speak your command: Once Alexa is activated, speak your command clearly and distinctly. You can ask Alexa to play a specific movie or TV show, search for content by genre or actor, and even control other smart home devices.

- Release the microphone button: Once you've finished speaking your command, release the microphone button. Alexa will respond to your command and take the appropriate action.

- Use natural language: Alexa understands natural language, so you can ask for content in a way that feels natural to you. For example, you can say "Play Stranger Things on Netflix" or "What are the new releases on Prime Video?"

- Control your smart home: Alexa can also be used to control other smart home devices such as lights, thermostats, and security cameras. You can ask Alexa to turn off the lights or adjust the temperature of your thermostat.

- Alexa Hands-Free: You can also enable Alexa Hands-Free on your Fire TV Stick 4K Max, which allows you to control your TV and other smart home devices without needing to press the microphone button on the remote.

'oice control with Alexa on the Fire TV Stick 4K Max makes it easy nd convenient to find and watch content and control other smart ome devices. With Alexa, you can use your voice to control your TV nd other devices, making the streaming experience more seamless nd enjoyable.

Iow to Connect Fire TV Stick to the TV

ou can connect Alexa to your Fire TV stick to control your TV and reaming content using voice commands. Here are the steps to llow:

- Make sure your Fire TV stick and Alexa device (such as a Echo Dot) are on the same Wi-Fi network.
- On your Fire TV stick, go to "Settings" -> "Device" -> "Alex Voice Remote."
- Select "Pair a new device."
- On your Alexa device, enable the "TV Control" skill in th Alexa app.
- Follow the on-screen instructions to link your Alexa device t your Fire TV stick.
- Once the devices are linked, you can use voice commands t control your Fire TV stick, such as "Alexa, play Stranger Thing on Fire TV" or "Alexa, turn off Fire TV".

Note: Make sure your Alexa device is compatible with your Fire T stick, also, you can use Alexa commands to control other devices th are connected to your Fire TV, such as soundbars or A/V receivers.

How to pair Fire Stick Remote with Alexa Echo

You can pair your Fire Stick remote with an Alexa Echo device control your Fire TV using voice commands. Here are the steps pair your Fire Stick remote with an Alexa Echo device:

- Make sure your Fire Stick and Alexa Echo device are on the same Wi-Fi network.
- Open the Alexa app on your smartphone or tablet.
- Tap on the menu icon (three horizontal lines) in the upper-left corner of the screen.
- Select "Settings."
- Select "TV & Video."
- Select "Fire TV."
- Select the Fire TV device you want to pair with your Alexa Echo device.
- Follow the on-screen instructions to complete the pairing process.
- Once the pairing is complete, you can use your Alexa Echo device to control your Fire TV by saying commands such as "Alexa, turn on my Fire TV" or "Alexa, play The Office on Fire TV."

Note: The pairing process may vary depending on the model of your Fire Stick and Alexa Echo device, please refer to the device's manual or the Alexa app's help center for specific instructions.

How to use fire tv with an echo speaker.

You can use your Echo speaker to control your Fire TV stick using voice commands. Here are the steps to follow:

- Make sure your Fire TV stick and Echo speaker are on the same Wi-Fi network.
- On your Echo speaker, open the Alexa app and go to "Settings" -> "TV & Video" -> "Fire TV"
- Select "Link a Device"
- Follow the on-screen instructions to link your Echo speaker to your Fire TV stick.
- Once the devices are linked, you can use voice commands to control your Fire TV stick through your Echo speaker, such as "Alexa, play Stranger Things on Fire TV" or "Alexa, turn off Fire TV".

You can also control the volume of your TV or soundbar, change channels, and navigate the Fire TV interface with commands like "Alexa, change the channel to NBC" or "Alexa, volume up on Fire TV"

Note: Make sure your Echo speaker is compatible with your Fire TV stick, also, you can use Alexa commands to control other devices that are connected to your Fire TV, such as soundbars or A/V receivers.

Usual Commands for Alexa

Alexa is a virtual assistant that can be used with a Fire TV stick to control your TV and streaming content using voice commands. Here are some general commands you can use:

- **Playback control**: "Play" / "Pause" / "Resume" / "Stop" / "Rewind" / "Fast forward."

- **Volume control**: "Volume up" / "Volume down" / "Mute" / "Unmute."

- **Channel control**: "Change the channel to (channel number or name)"

- **Navigation**: "Go back" / "Go home" / "Open (app name)"

- **Search**: "Search for (movie or TV show title)"

- **Information**: "What's on (channel name)?" / "What's the weather like?" / "What's the news?"

- **Control smart devices**: "Turn on the living room lights" / "Set the thermostat to 72 degrees"

- **Timers and alarms**: "Set a timer for 20 minutes" / "Wake me up at 7am tomorrow."

Note: These commands are general commands, some might not work for certain devices or apps. Also, some commands might have variations depending on the device or service you're using, you can also check the Alexa App or the Alexa Skills store for more commands and features.

Chapter 8: Additional Connections and Apps

How to Connect the Fire TV Stick to Multiple TV

You can connect your Fire TV stick to multiple TVs, but you will need to register the stick to each TV separately. Here are the steps to follow:

- Make sure your Fire TV stick is unregistered from the current TV. To do this, go to "Settings" -> "Device" -> "About" -> "Reset to Factory Defaults" -> "Reset"

- Connect the Fire TV stick to the second TV and follow the on-screen instructions to set up the device.

- Once the Fire TV stick is set up, go to "Settings" -> "Device" -> "Developer options" and turn on "ADB debugging" and "Apps from Unknown Sources"

- Connect your Fire TV stick to the first TV and follow the on-screen instructions to set up the device.

- Repeat the process for any additional TVs you want to connect the Fire TV stick to.

Note: The Fire TV stick is designed to be used in one location at a time, so you will have to switch the HDMI input on the TV to use the ire TV stick on a different TV. Additionally, you will have to log in ıd out of your Amazon account each time you switch TVs.

low to Use the Fire Stick on a Monitor

ou can use a Fire TV stick on a monitor by connecting it to the ıonitor's HDMI port. Here are the steps to follow:

- Turn off the monitor and connect the Fire TV stick to the HDMI port using an HDMI cable.
- Turn on the monitor and set the input to the HDMI port that the Fire TV stick is connected to.
- Once the monitor detects the Fire TV stick, the setup screen will appear. Follow the on-screen instructions to set up your Fire TV stick.
- Once the Fire TV stick is set up, you can use it to access streaming services, play games, and more.
- To navigate the interface, you can use the Fire TV remote or connect a Bluetooth keyboard or mouse to the Fire TV stick.

ote: Some monitors may not support audio output through the DMI port, in this case, you will need to connect a separate audio ·vice, such as speakers or a soundbar, to the Fire TV stick. dditionally, some monitors may have a lower resolution than a pical TV, this may affect the quality of the video.

How to Use Bluetooth on Fire TV

Bluetooth is a wireless technology that allows you to connect device such as headphones, speakers, and game controllers to your Fire T stick. Here are the steps to use Bluetooth on your Fire TV stick:

- Turn on your Fire TV stick and navigate to the home screen.
- Select the "Settings" option located on the top of the screen.
- Select "Controllers and Bluetooth Devices"
- Select "Other Bluetooth Devices"
- Put your Bluetooth device in pairing mode, usually by holdin down a button on the device until a light starts flashing.
- Select the device you want to connect to from the list available devices.
- Follow the on-screen instructions to complete the pairin process.
- Once connected, your Fire TV stick will automatically conne to the device whenever it is in range and turned on.

Note: Some Fire TV stick models do not have built-in Bluetoo capability, in that case, you may need to purchase a separa Bluetooth adapter to connect to other devices. Additionally, th process of pairing and connecting devices may vary depending on th device and the Fire TV stick model you are using.

Screen Mirroring

Screen mirroring is a feature that allows users to mirror their smartphone, tablet, or PC screen on their TV. This feature is available on the Amazon Fire TV Stick 4K Max and can be used to share content or play mobile games on a bigger screen. Here's a detailed look at how to use screen mirroring on the Fire TV Stick 4K Max:

- Enable screen mirroring on your Fire TV Stick 4K Max: To enable screen mirroring, go to the settings menu on the Fire TV Stick 4K Max home screen and select "Display & Sounds" and then "Enable Display Mirroring."

- Enable screen mirroring on your device: On your smartphone, tablet, or PC, go to the settings menu and look for the "screen mirroring" or "smart view" option.

- Connect your device to your Fire TV Stick 4K Max: Once screen mirroring is enabled on both devices, your Fire TV Stick 4K Max should appear as a device to connect to on your mobile device or PC. Select your Fire TV Stick 4K Max and the device will connect to it.

- Start mirroring: Once connected, your device's screen will appear on your TV. You can use your device as normal, and whatever you do on your device will be displayed on your TV.

It's worth mentioning that the quality of the mirrored content may vary depending on the device and the streaming quality. Also, some applications may not work properly when mirrored.

Overall, screen mirroring is a useful feature that allows users to share content or play mobile games on a bigger screen. With the Fire TV Stick 4K Max, users can easily enable screen mirroring and connect their mobile device or PC to the TV for a seamless experience.

Chapter 9: Remote Repair and Replacement

How to repair the remote

If your Fire TV stick remote is not working properly, there are several troubleshooting steps you can take to repair it. Here are some common solutions to fix a malfunctioning Fire TV stick remote:

- Check the batteries: Make sure the batteries in the remote are fresh and inserted correctly.

- Check the connection: Make sure the remote is within range of the Fire TV stick and there are no obstacles blocking the signal.

- Restart the remote: Remove the batteries from the remote, wait for a few seconds, and then reinsert the batteries.

- Restart the Fire TV stick: Unplug the Fire TV stick from the power source, wait for a few seconds, and then plug it back in.

- Check for software updates: Go to the Settings menu on your Fire TV stick, select "Device," and then select "About." Check for any available software updates and install them.

- Pair the remote again: Go to the Settings menu on your Fire TV stick, select "Device," and then select "Bluetooth." Select "Amazon Fire TV Remote" and then press and hold the home button on the remote until the remote is paired.

- Get a replacement: If none of the above solutions work, you may need to get a replacement remote. You can purchase a replacement remote from Amazon or contact Amazon customer service for assistance.

Note: It's important to note that these solutions are general troubleshooting steps and may not apply to all cases. Additionally, some issues may be caused by hardware defects and in those cases, a replacement remote may be the only solution.

Replacing the remotes

If your Fire TV stick remote is lost, broken, or malfunctioning, you can replace it with a new one. Here are the steps to replace your Fire TV stick remote:

- Purchase a new Fire TV stick remote: You can purchase a new remote from Amazon or other retailers that sell Fire TV accessories.

- Pair the new remote with your Fire TV stick: Insert the batteries in the new remote and press and hold the home button on the remote for about 10 seconds, until the remote is paired with your Fire TV stick.

- Check the settings: Go to the settings menu on your Fire TV stick, select "Device," and then select "Bluetooth." Select "Amazon Fire TV Remote" and then press and hold the home button on the remote until the remote is paired.

- Test the remote: Once the remote is paired, test the remote by navigating through the Fire TV stick's menus and checking if all the buttons are working properly.

Alternatively, you can also use the fire tv remote app on your smartphone as a replacement for a lost or broken remote.

Note: It's important to note that the replacement remote should be of the same model and type of your original remote, to ensure compatibility and full functionality. Additionally, it's recommended to check the warranty of your device, as some manufacturers provide replacement remote services.

ow to restart with the remote

here are a few different ways to restart your Fire TV stick using the
emote control. Here are the steps to restart your Fire TV stick using
e remote:

- Press and hold the select button and the play/pause button at the same time for about five seconds. The Fire TV stick will restart and the Amazon logo will appear on the screen.

- Press and hold the Home button on your remote for at least 10 seconds. The Fire TV stick will restart and the Amazon logo will appear on the screen.

- Go to the settings option on the top of the screen, navigate to device and select "Restart"

ote: Restarting your Fire TV stick will close all open apps and
turn you to the home screen, but it will not delete any of your
rsonal settings or data. Additionally, It's important to note that these
eps may vary depending on the specific model and software version
your Fire TV stick.

Chapter 10: Customizing Your Fire TV Stick 4K Max

Customizing your Amazon Fire TV Stick 4K Max allows you t personalize your experience and make the device work best for yo Here are some of the ways you can customize your Fire TV Stick 4 Max:

- **Customize the home screen**: You can customize the hom screen by adding or removing apps, changing the layout, ar rearranging the order of the items on the home screen. T customize the home screen, go to the settings menu and sele "Customize Your Home Screen."

- **Set parental controls**: You can set parental controls on yo Fire TV Stick 4K Max to restrict access to certain content ar apps. To set parental controls, go to the settings menu and sele "Parental Controls."

- **Manage subscriptions**: You can manage your subscriptions streaming services such as Netflix, Prime Video, and Disney on your Fire TV Stick 4K Max. To manage subscriptions, go the settings menu and select "Subscriptions."

- **Change the theme**: You can change the theme of your Fire TV Stick 4K Max to match your preference. To change the theme, go to the settings menu and select "Display & Sounds" and then "Change Theme."

- **Change the language**: You can change the language of your Fire TV Stick 4K Max to match your preference. To change the language, go to the settings menu and select "Device" and then "Device Options" and then "Language."

- **Change the Display Resolution**: You can also change the display resolution to match your TV's capability. To change the resolution, go to the settings menu and select "Display & Sounds" and then "Change Resolution."

By customizing your Fire TV Stick 4K Max, you can make the device work best for you and your family. Whether it's setting parental controls or managing subscriptions, the Fire TV Stick 4K Max offers a variety of customization options to make your streaming experience even better.

How to enable 4k playback

To enable 4K playback on a Fire TV stick, you need to ensure that your TV and Fire TV stick support 4K resolution, and that the content you are trying to play is available in 4K. Here are the steps to follow:

- Check that your TV supports 4K resolution. You can check this in the TV's manual or by going to the TV's settings menu.

- Check that your Fire TV stick supports 4K resolution. You can check this in the Fire TV stick's manual or by going to the Fire TV stick's settings menu.

- Connect your Fire TV stick to your TV using a High-Speed HDMI cable that supports 4K resolution

- Go to "Settings" -> "Display & Sounds" -> "Video Resolution"

- Select "4K"

- Verify that the 4K resolution is now selected.

- Search for 4K content and play it, if the content is not 4k, the Fire TV will automatically adjust the resolution accordingly.

Note: Make sure that your internet connection is fast enough to support 4K streaming, as it requires a minimum of 15Mbps. Also, some content providers may not offer 4K streaming in your region or may charge extra for it.

Changing the Background

One of the ways to customize your Amazon Fire TV Stick 4K Max is by changing the background or wallpaper of the home screen. Here's a detailed look at how to change the background on your Fire TV Stick 4K Max:

- Go to the settings menu: From the home screen, navigate to the settings menu by using the remote's directional buttons and press the center button to select "Settings."

- Select "Display & Sounds": From the settings menu, navigate to "Display & Sounds" and press the center button to select it.

- Select "Change the Background": Once you're in the Display & Sounds section, you'll see an option for "Change the Background", press the center button to select it.

- Select your desired background: You'll see a variety of backgrounds to choose from, including solid colors, patterns, and pre-loaded wallpapers. Use the remote's directional buttons to navigate through the options and press the center button to select your desired background.

- Confirm your selection: Once you've selected your desired background, you'll be prompted to confirm your selection. Press the center button to confirm and your new background will be applied.

It's worth mentioning that, on some versions of the Fire TV Stick, you may not have the option to change the background, or it may be under a different name.

By changing the background on your Fire TV Stick 4K Max, you can personalize your device and make it more visually appealing. The option to change the background is a simple but effective way to customize your device and make it feel like your own.

How to Turn off the Click Sound on the Fire TV Stick

o turn off the click sound on your Fire TV stick, you can follow ıese steps:

- Go to the settings option on the top of the screen.
- Navigate to the "Device" option and select "Accessibility."
- Scroll down to the "Navigation" section and select "Click Sound."
- Change the toggle switch to "Off."

lternatively, you can also try disabling the click sound from the mote by going to the settings in the remote app.

ote: The exact steps and location of the option to turn off the click und may vary depending on the specific model and software ersion of your Fire TV stick. Additionally, disabling the click sound ay affect the functionality or performance of your Fire TV stick.

Chapter 11: Best Apps for the Fire TV Stick

There are many apps available on the Fire TV stick that can enhanc your streaming experience, but some of the most popular app include:

- **Netflix**: One of the most popular streaming services, Netfl offers a wide variety of TV shows, movies, and original conten
- **Amazon Prime Video**: Amazon's streaming service offers wide variety of TV shows, movies, and original content, as we as access to live sports and events.
- **Hulu**: With a wide variety of TV shows, movies and origin content, Hulu is a great app for streaming TV shows ar movies.
- **YouTube**: Allows you to watch videos, music, and live strean on your TV.
- **Disney+**: Offers a wide variety of movies and TV shows fro Disney, Marvel, Star Wars, and more.
- **HBO Max**: Offers a wide variety of movies and TV show including HBO original content, as well as classic shows a movies.
- **Plex**: Allows you to stream your own personal media librar including movies, TV shows, music, and photos from yo computer or mobile device to your TV.
- **Twitch**: Allows you to watch live streams of video games a other content from around the world.

These apps are some of the most popular and highly rated on the Fire TV stick. It is worth noting that availability of apps may vary depending on your location and device.

How to Use Side Loading Apps

Side loading is the process of installing apps on your Fire TV stick that are not available on the Amazon Appstore. This can be done by using an Android Debug Bridge (ADB) tool on your computer and connecting it to your Fire TV stick. Here are the steps to side load apps on your Fire TV stick:

- Make sure your Fire TV stick and computer are connected to the same Wi-Fi network.
- On your computer, install the ADB tool and the necessary drivers for your Fire TV stick.
- On your Fire TV stick, go to Settings, select "My Fire TV," and turn on "ADB Debugging."
- Connect your computer and Fire TV stick by entering the IP address of your Fire TV stick in the ADB tool.
- Once connected, use the ADB tool to install the app you want to side load on your Fire TV stick.
- After the installation is complete, the app will be listed on the home screen and can be launched like any other app.

Note: Side loading apps on your Fire TV stick is a process that requires technical knowledge and may not be suitable for all users.

Additionally, side loading apps can be risky as it can void your warranty, cause instability and conflicts with other apps on your device. It's also important to note that not all apps will work correctly and some apps may not be legal to use in some countries.

Amazon Prime Video

Amazon Prime Video is a streaming service provided by Amazon that offers a wide variety of TV shows, movies, and original content for users to watch on-demand. It is available to users who have an Amazon Prime membership, which also includes other benefits such as free two-day shipping, access to the Kindle Lending Library, and more. Some of the features of Amazon Prime Video include:

- **A vast library of TV shows and movies**: Amazon Prime Video offers a wide range of content including new and popular movies, TV shows, and original content such as The Marvelous Mrs. Maisel, Jack Ryan, and The Man in the High Castle.

- **Access to Amazon Prime Video Channels**: Users can also subscribe to additional channels such as HBO, STARZ, and more, as an add-on to their Prime membership.

- **Available on multiple devices**: Amazon Prime Video can be accessed on a variety of devices including Fire TV sticks, smart TVs, gaming consoles, smartphones, tablets and more.

- **Offline viewing**: Users can download TV shows and movies to watch offline, which is especially useful for traveling or when internet connection is not available.

- **Parental controls**: Users can set up a PIN and content filters to restrict access to certain content and features on Amazon Prime Video.

- **Multiple profiles**: Users can create and manage multiple profiles on Amazon Prime Video, which allows different members of the household to have their own watchlist and recommendations.

Note: Amazon Prime Video is a separate service from Amazon Prime, and requires a subscription to access the content. Some of the content on the service is exclusive to Amazon Prime Video and is not available on other streaming services.

How to Subscribe to Amazon Prime Membership

Amazon Prime is a subscription service provided by Amazon that offers a wide variety of benefits, including free two-day shipping, access to the Kindle Lending Library, and access to the streaming service Amazon Prime Video. Here are the steps to subscribe to Amazon Prime:

- Go to the Amazon website and sign in to your account.
- Go to the "Your Account" page and select "Prime."
- Click on the "Start your 30-day free trial" button.
- Enter your payment information and click "Start your free trial."
- Once the free trial is over, Amazon Prime will automatically renew at a monthly or annual rate, depending on the plan you selected.
-

Alternatively, you can also subscribe to Amazon Prime from the Fire TV Stick:

- Go to the home screen and navigate to the “Apps” section.
- Scroll down to the “Amazon Prime Video” app and select it
- Select the “Start your free trial” button.

- Follow the on-screen instructions to complete the subscription process.

Iote: Amazon Prime membership also includes access to a wide ınge of benefits such as Prime Music, Prime Reading, and more. .mazon Prime membership can be shared with other members of the ousehold using Amazon Household feature, which allows multiple sers to share the benefits of the subscription.

Vatching Youtube

- YouTube is a popular video-sharing platform that can be accessed on your Fire TV stick. Here are the steps to watch YouTube on your Fire TV stick:

- Turn on your Fire TV stick and navigate to the home screen.
- Select the "Apps" option located on the top of the screen.
- Scroll down to the "YouTube" app and select it.
- Sign in to your Google account (if you haven't already)
- Browse the available videos or use the search bar to find specific content.
- Select the video you want to watch and press the "Play" button on your remote control to start the playback.

lternatively, you can also use the voice search feature of your Fire V stick to search for videos on YouTube:

- Press the microphone button on your Alexa Voice Remote an say "YouTube."
- Say the name of the video or channel you want to watch.
- Select the video you want to watch and press the "Play" buttc on your remote control to start the playback.

Note: Some YouTube videos may be blocked in certain regions c may require a YouTube Premium subscription to watch. Additionall YouTube content may not be suitable for all ages, and parent controls are recommended.

Watching Hbo Max

HBO Max is a streaming service provided by HBO that offers a wic variety of TV shows, movies, and original content for users to watc on-demand. Here are the steps to watch HBO Max on your Fire T stick:

- Turn on your Fire TV stick and navigate to the home screen.
- Select the "Apps" option located on the top of the screen.
- Scroll down to the "HBO Max" app and select it.
- If you don't have the app already, you will be prompted download and install it.
- Sign in to your HBO Max account (if you haven't already)
- Browse the available shows and movies or use the search bar find specific content.
- Select the show or movie you want to watch and press tl "Play" button on your remote control to start the playback.

Alternatively, you can also use the voice search feature of your Fire TV stick to search for shows and movies on HBO Max:

- Press the microphone button on your Alexa Voice Remote and say "HBO Max."
- Say the name of the show or movie you want to watch.
- Select the show or movie you want to watch and press the "Play" button on your remote control to start the playback.

Note: To access HBO Max, you will need to have a subscription, which can be purchased directly through the app or through a participating cable or streaming provider. Additionally, some content may not be available in all regions and may require a VPN.

How to Install Apps on a Fire Stick

Installing apps on a Fire TV stick is a simple process. Here are the steps to follow:

- Turn on your Fire TV stick and navigate to the home screen.
- Using the remote, select the "Search" option located on the top of the screen.
- Type in the name of the app you want to install.
- Select the app from the search results and then select "Get" or "Download" to begin the installation process.
- Once the app is downloaded and installed, select "Open" to launch the app.

Alternatively, you can also install apps by going to the "Apps" section on the home screen, then selecting "Categories" and selecting the app from the list.

If the app you want to install is not available on the Amazon Appstore, you can also use the "Downloader" app to sideload apps from other sources.

Note: Some apps may require a subscription or additional fee to use. Also, make sure your FireTV stick is connected to the internet and has enough storage for the app to install.

Access to apps and games: The Fire TV allows users to access a wide variety of apps and games, including popular streaming services, social media platforms, and gaming apps.

Uninstalling Apps using Alexa

You can uninstall apps from your Fire TV using Alexa voice commands. Here are the steps to follow:

- Press and hold the microphone button on your Alexa Voice Remote or use an Alexa-enabled device such as an Echo Dot.
- Say "Alexa, open the Alexa app" to launch the Alexa app on your Fire TV.
- Select the "Devices" option from the bottom menu.
- Select your Fire TV from the list of devices.
- Select "Manage installed apps" from the options.
- Select the app you want to uninstall.
- Select "Uninstall" to remove the app from your Fire TV.

Note: Some apps may not be able to be uninstalled using Alexa, and you will have to use the FireTV remote instead to remove them. Also, uninstalling an app will delete all the data and settings associated with the app.

Managing Apps on Fire TV Stick

Managing apps on a Fire TV stick is a simple process. Here are the steps to follow:

- Turn on your Fire TV stick and navigate to the home screen.
- Using the remote, select the "Settings" option located on the top of the screen.
- Select "Applications"
- Here you have several options to manage your apps, such as:
- Manage Installed Applications: This option allows you to view all the apps currently installed on your Fire TV stick. You can

select an app to view more information about it or choose to "Force Stop" or "Uninstall" it.

- Available for Download: This option shows you all the apps that are available for download from the Amazon Appstore.

- Downloaded: This option shows you all the apps that you have downloaded on your Fire TV stick but haven't yet installed.

- Cloud: This option shows you all the apps that you have downloaded on other devices and are available to download on your Fire TV stick.

- You can also filter the apps by categories, like "Games", "Utility" or "Entertainment."

- Additionally, you can control the automatic updates of your apps, by going to "Settings" -> "Applications" -> "Auto-Update Apps" and choose if you want the apps to update automatically, manually or never.

Note: Some apps may require a subscription or additional fee to use. Also, make sure your FireTV stick is connected to the internet and has enough storage for the app.

Uninstalling Apps from Fire TV Stick

ere are the steps to uninstall an app from your Fire TV stick:

- Turn on your Fire TV stick and navigate to the home screen.
- Using the remote, select the "Settings" option located on the top of the screen.
- Select "Applications"
- Select "Manage Installed Applications"
- Select the app you want to uninstall.
- Select "Uninstall" to remove the app from your Fire TV stick.
- Confirm the uninstallation by selecting "Uninstall" again.

ote: Some apps may not be able to be uninstalled, and they will have be disabled instead. Also, uninstalling an app will delete all the ata and settings associated with the app. Additionally, uninstalling an pp may not cancel any active subscriptions or recurring charges lated to the app.

How to Clear the Cache on Fire TV Stick

Clearing the cache on a Fire TV stick can help improve its performance and fix any issues that you may be experiencing with certain apps or streaming content. Here are the steps to follow:

- Turn on your Fire TV stick and navigate to the home screen.
- Using the remote, select the "Settings" option located on the to of the screen.
- Select "Applications"
- Select "Manage Installed Applications"
- Select the app for which you want to clear the cache.
- Select "Clear cache" to remove the cached data.
- Confirm the action by selecting "OK."

Note: Clearing the cache will not delete any of your personal data or settings associated with the app, it will only delete temporary files that may be causing issues. Additionally, Clearing the cache will not fix all the issues, but it may help with some performance problem. If you continue to have problems with an app, you may want to try forc stopping the app or reinstalling it.

How to hide uninstalled Apps.

If you have uninstalled apps from your Fire TV stick that you no longer want to see in the app list, you can hide them from view. Here are the steps to follow:

- Turn on your Fire TV stick and navigate to the home screen.
- Using the remote, select the "Settings" option located on the top of the screen.
- Select "Applications"
- Select "Manage Installed Applications"
- Select the app you want to hide.
- Select "Hide from recent"
- Confirm the action by selecting "OK"
- This will remove the app from the list of recently used apps, so it will no longer appear when you scroll through the apps on your Fire TV.

Note: Hiding an app will not completely remove it from your Fire TV, it will still be installed and taking up storage. If you want to completely remove an app, you will have to uninstall it. Also, hiding an app will not cancel any active subscriptions or recurring charges related to the app.

Chapter 12: Setting Parental Controls

The Amazon Fire TV Stick 4K Max allows you to set parental controls to restrict access to certain content and apps. This feature is useful for parents who want to ensure their children are not exposed to inappropriate content. Here's a detailed look at how to set parental controls on the Fire TV Stick 4K Max:

- Go to the settings menu: From the home screen, navigate to the settings menu by using the remote's directional buttons and press the center button to select "Settings."

- Select "Parental Controls": From the settings menu, navigate to "Parental Controls" and press the center button to select it.

- Create a PIN: You will be prompted to create a PIN to use for parental controls. Use the on-screen keyboard to enter a four-digit PIN and press the center button to confirm.

- Set content restrictions: Once you've set a PIN, you can set content restrictions by selecting the "Content Restrictions" option. You can choose to restrict access to specific TV shows, movies, and apps based on their rating or type.

- Set purchase restrictions: You can also set purchase restrictions by selecting the "Purchase Restrictions" option. This allows you to restrict the ability to purchase or rent content on the Fire TV Stick 4K Max based on the content's rating or type.

- Manage profiles: You can also create multiple profiles for different users and set different restrictions for each profile.

It's worth mentioning that, if you forget your PIN, you can reset it by selecting the "Forgot PIN" option on the Parental Controls screen and following the prompts.

By setting parental controls on your Fire TV Stick 4K Max, you can ensure that your children are not exposed to inappropriate content. The feature allows you to set content and purchase restrictions, create multiple profiles, and manage them with a PIN, providing parents with peace of mind and control over their children's viewing experience.

How to Change Parental Control Pin

You can change the Parental Control PIN on your Fire TV stick to restrict access to certain content and features. Here are the steps to follow:

- Turn on your Fire TV stick and navigate to the home screen.
- Select the "Settings" option located on the top of the screen.
- Select "Preferences"
- Select "Parental Controls"
- Enter your current Parental Control PIN and select "OK."
- Select "Change PIN"
- Enter your new Parental Control PIN and select "OK."
- Re-enter your new Parental Control PIN to confirm and select "OK."

Note: Make sure to choose a strong and unique PIN that is easy for you to remember but difficult for others to guess. Also, If you forget our Parental Control PIN, you will have to reset your FireTV to actory settings to change it.

Iow to Block Explicit Songs

ou can block explicit songs on a Fire TV stick by enabling the arental Control feature and setting a content filter. Here are the steps follow:

- Turn on your Fire TV stick and navigate to the home screen.
- Select the "Settings" option located on the top of the screen.
- Select "Preferences"
- Select "Parental Controls"
- Enable the Parental Control feature and set a PIN.
- Select "Music"
- Select "Filter explicit songs."
- Confirm your PIN.
- This will block explicit songs on all music streaming services on your Fire TV stick.

Note: Some streaming services may not have explicit content filter, i this case, you will have to use the filter of the specific service. Als this feature may not block all explicit content, as some songs may n be labeled as explicit by the streaming service or may not be in th Fire TV database of explicit songs. Additionally, you can use thi feature to block explicit content on other apps, such as movies, T shows and games.

How to set up Amazon FreeTime on a Fire Stick

Amazon FreeTime is a parental control feature on the Fire TV stic that allows you to set limits on content and screen time for kids. He are the steps to set it up:

- Turn on your Fire TV stick and navigate to the home screen.
- Select the "Settings" option located on the top of the screen.
- Select "Preferences"
- Select "Parental Controls"
- Enable the Parental Control feature and set a PIN.
- Select "Amazon FreeTime"
- Follow the on-screen instructions to set up Amazon FreeTime.
- You can choose to set limits on screen time and content, s educational goals, and create a personalized profile for ea child in your household.

- Once set up, the child will have access to a customized home screen with age-appropriate content and features.

Note: To use Amazon FreeTime, you will need an Amazon account and you may also need to subscribe to Amazon Kids+ (formerly known as FreeTime Unlimited) which is a paid service that gives access to a wide range of kid-friendly content, including books, videos, games, and educational apps. Additionally, you can use the Amazon FreeTime feature on other Amazon devices, such as the Kindle, tablets, and phones.

Write about how to subscribe to Amazon Kids+

Amazon Kids+ (formerly known as FreeTime Unlimited) is a subscription service that provides access to a wide range of kid-friendly content, including books, videos, games, and educational apps. Here are the steps to subscribe:

- Turn on your Fire TV stick and navigate to the home screen.
- Select the "Settings" option located on the top of the screen.
- Select "Preferences"
- Select "Parental Controls"
- Select "Amazon FreeTime"
- Select "Subscribe to Amazon Kids+"
- Follow the on-screen instructions to subscribe, including providing payment information.

- Once subscribed, you will have access to the full range of content on Amazon Kids+, including books, videos, games, and educational apps that are appropriate for children.

Note: You can also subscribe to Amazon Kids+ on other devices, such as Kindle, tablets, and phones. Amazon Kids+ is designed for children aged 3-12, and you can create a personalized profile for each child in your household. Additionally, you can manage your subscription and cancel it at any time.

Chapter 13: Managing Your Subscriptions

The Amazon Fire TV Stick 4K Max allows you to manage your subscriptions to streaming services such as Netflix, Prime Video, Disney+ and others. This feature is useful for keeping track of your subscriptions, canceling or modifying them, and also for discovering new content. Here's a detailed look at how to manage your subscriptions on the Fire TV Stick 4K Max:

- Go to the settings menu: From the home screen, navigate to the settings menu by using the remote's directional buttons and press the center button to select "Settings."

- Select "Subscriptions": From the settings menu, navigate to "Subscriptions" and press the center button to select it.

- View your subscriptions: Once you're in the Subscriptions section, you'll see a list of your current subscriptions. You can use the remote's directional buttons to navigate through the list and press the center button to select a subscription.

- Modify or cancel a subscription: Once you've selected a subscription, you'll have the option to modify or cancel it. Depending on the service, you may have the option to change your plan, update your payment method, or cancel the subscription.

- Discover new content: You can also discover new content by browsing different streaming services and their available plans. You can use the remote's directional buttons to navigate through the options and press the center button to select a service.

It's worth mentioning that some services may not be available on all regions and also, the options available to modify or cancel a subscription may vary depending on the service.

By managing your subscriptions on your Fire TV Stick 4K Max, you can keep track of your subscriptions, cancel or modify them, and discover new content. This feature is useful for keeping your subscription costs under control and discovering new content.

How to create subscriptions

To subscribe to a service on a Fire TV stick, you can follow these steps:

- Turn on your Fire TV stick and navigate to the home screen.
- Find the service or app that you want to subscribe to and select it.
- Look for a "Subscribe" or "Start your free trial" option on the app's main menu or within the app's settings.
- Select the subscription option and follow the on-screen instructions to complete the subscription process.

- You will be prompted to enter your payment information and confirm the subscription.
- Once the subscription is complete, you will have access to the content or features included in the subscription.

Note: Subscriptions will automatically renew on a monthly or annual basis, and you will be charged accordingly. You can manage your subscriptions by going to the "Settings" -> "Accounts & Lists" -> "Your Subscriptions" in the Fire TV menu, where you can cancel or modify your subscriptions. Also, some apps or services may offer a free trial period before you have to subscribe.

How to Create Channel and App Subscriptions

There are several ways to make channel and app subscriptions on a Fire TV stick:

- From the home screen: Navigate to the home screen and select the "Apps" or "Channels" option. Browse through the available channels or apps and select the one you want to subscribe to. Look for a "Subscribe" or "Start your free trial" option on the app's main menu or within the app's settings. Select the subscription option and follow the on-screen instructions t complete the subscription process.

- Using Alexa: Press the microphone button on your Alexa Voic Remote and say "subscribe to (channel or app name)" or "start free trial of (channel or app name)". Alexa will guide yo through the process.

- Using the Amazon Appstore: You can also browse an subscribe to channels and apps through the Amazon Appstor Navigate to the Appstore on your Fire TV stick, search for th channel or app you want to subscribe to, and select th "Subscribe" or "Start your free trial" option. Follow the on screen instructions to complete the subscription process.

Note: Subscriptions will automatically renew on a monthly or annu basis, and you will be charged accordingly. You can manage yo subscriptions by going to the "Settings" -> "Accounts & Lists" - "Your Subscriptions" in the Fire TV menu, where you can cancel o modify your subscriptions. Some apps or services may offer a fre trial period before you have to subscribe.

How to Create Prime Video Channel Subscription

Prime Video Channels is a feature of Amazon Prime Video th allows you to subscribe to additional channels, such as HBO, STAR and more. Here are the steps to create a Prime Video Chann subscription:

- Go to the Amazon website, and sign in to your account.

- Go to the "Your Account" page and select "Prime Video Channels."
- Browse through the available channels and select the one you want to subscribe to.
- Select the "Subscribe" button and follow the on-screen instructions to complete the subscription process.

Alternatively,

- Open the Amazon Prime Video app on your Fire TV stick.
- Go to the "Settings" menu.
- Select "Prime Video Channels"
- Browse through the available channels and select the one you want to subscribe to.
- Select the "Subscribe" button and follow the on-screen instructions to complete the subscription process.

Note: Some channels may offer a free trial before you have to subscribe. Once you complete the subscription process, you'll be able to watch the content provided by that channel on your Fire TV stick. If you have any trouble subscribing or don't see your desired channel you can also check the availability of channels in your region, as well as the pricing, conditions, and terms of service that may vary depending on the channel provider.

How to Cancel Prime Video Channel Subscription

- Here are the steps to cancel a Prime Video Channel subscription:
- Go to the Amazon website, and sign in to your account.
- Go to the "Your Account" page and select "Prime Video Channels."
- Select the channel you want to cancel and choose "Cancel Channel."

- Confirm the cancellation by selecting "Cancel Channel."
-

Alternatively,

- Open the Amazon Prime Video app on your Fire TV stick.
- Go to the "Settings" menu.
- Select "Prime Video Channels"
- Select the channel you want to cancel and choose "Cancel Channel."
- Confirm the cancellation by selecting "Cancel Channel."

Note: Canceling a channel may not cancel your subscription immediately, you may be able to access the content until the end of the current billing period. Additionally, some channels may have their own subscription management process, so you will need to check with the channel provider for more information on how to cancel a subscription.

How to Create and Cancel Streaming Service App Subscriptions

Many streaming services, such as Netflix, Hulu, and Disney+, offer the option to subscribe to their service through their app on your Fire TV stick. Here are the general steps to create and cancel a streaming service app subscription on your Fire TV stick:

Creating a subscription:

- Open the app for the streaming service you want to subscribe to on your Fire TV stick.
- Look for a "Start Free Trial" or "Subscribe" button and select it.
- Follow the on-screen instructions to complete the subscription process, which may include providing payment information.

Canceling a subscription:

- Open the app for the streaming service you want to cancel the subscription for on your Fire TV stick.
- Go to the app's settings or account section, usually located in the menu or profile section.
- Look for a "Cancel Subscription" or "Manage Subscription" option and select it.
- Confirm the cancellation by following the on-screen instructions.

Note: Canceling a subscription may not cancel it immediately, you may be able to access the content until the end of the current billing period. Additionally, the process of creating or canceling a subscription may vary depending on the service and you may need to check the service's website or contact customer service for further instructions.

Chapter 14: Troubleshooting

Despite being a reliable device, the Amazon Fire TV Stick 4K Max may experience issues from time to time. Here are some common issues and troubleshooting steps to resolve them:

- **No video or audio**: If you're not getting any video or audio when streaming content, check the connections on your TV and Fire TV Stick 4K Max to make sure they're securely plugged in. Also, check the resolution settings on your TV and make sure they match the resolution of your Fire TV Stick 4K Max.

- **Remote not working**: If your remote is not working, check the batteries and make sure they're properly inserted. You can also try resetting the remote by removing the batteries and reinserting them. If the remote still doesn't work, contact Amazon's customer service for assistance.

- **Content not loading**: If the content you're trying to stream is not loading, check your internet connection and make sure it's stable. Also, check to make sure your Fire TV Stick 4K Max software is up-to-date.

- **Issues with Alexa**: If you're having issues with Alexa, check the microphone on your remote and make sure it's not blocked. Also, check to make sure your Fire TV Stick 4K Max software is up-to-date and that Alexa is enabled in the settings.

- **Slow Performance**: If your Fire TV Stick 4K Max is running slowly, try closing any apps that are running in the background, and also check the available storage space and delete any unnecessary files.

- **General troubleshooting**: If you're experiencing an issue that's not listed above, try restarting your Fire TV Stick 4K Max by unplugging it from the power source and plugging it back in. This can resolve many issues.

's worth mentioning that, Amazon customer service is available 24/7) help with any troubleshoot you may have.

y following these troubleshooting steps, you can resolve most ommon issues with your Amazon Fire TV Stick 4K Max. If the roblem persists, contact Amazon customer service for assistance.

ommon Issues and Solutions

he Amazon Fire TV Stick 4K Max is a reliable device, but it may xperience issues from time to time. Here are some common issues ıd solutions for them:

- **No video or audio**: If you're not getting any video or audio when streaming content, check the connections on your TV and Fire TV Stick 4K Max to make sure they're securely plugged in. Also, check the resolution settings on your TV and make sure they match the resolution of your Fire TV Stick 4K Max.

- **Remote not working**: If your remote is not working, check the batteries and make sure they're properly inserted. You can also try resetting the remote by removing the batteries and reinserting them. If the remote still doesn't work, contact Amazon's customer service for assistance.

- **Content not loading**: If the content you're trying to stream is not loading, check your internet connection and make sure it's stable. Also, check to make sure your Fire TV Stick 4K Max software is up-to-date.

- **Issues with Alexa**: If you're having issues with Alexa, check the microphone on your remote and make sure it's not blocked. Also, check to make sure your Fire TV Stick 4K Max software is up-to-date and that Alexa is enabled in the settings.

- **Slow Performance**: If your Fire TV Stick 4K Max is running slowly, try closing any apps that are running in the background, and also check the available storage space and delete any unnecessary files.

- **Buffering**: Buffering is caused by a slow internet connection. To fix it, you can try resetting your router, moving your device closer to the router, or upgrading your internet plan.

- **Subscription issues**: Subscription issues can be caused by a problem with the payment method, a problem with the subscription itself, or by a problem with the account. To fix it, check your payment method, check the status of the subscription, and contact customer service if necessary.

- **Black screen**: This issue can be caused by a problem with the HDMI connection, a problem with the TV settings, or a problem with the Fire TV Stick 4K Max. To fix it, check the HDMI connection, check the TV settings, and restart the Fire TV Stick 4K Max.

- **Error messages**: Error messages can indicate a problem with the app, the internet connection, or the Fire TV Stick 4K Max. To fix it, check the internet connection, close and reopen the app, or restart the Fire TV Stick 4K Max.

- **Connectivity issues**: Connectivity issues can be caused by a problem with the router, the internet connection, or the Fire TV Stick 4K Max. To fix it, check the router, check the internet connection, and restart the Fire TV Stick 4K Max.

It's worth mentioning that, not all solutions will work for all issues, and that some issues may require contacting customer service for assistance.

By understanding and troubleshooting these common issues, you can quickly resolve problems with your Amazon Fire TV Stick 4K Max and continue to enjoy your streaming experience. If the problem persists, contact Amazon customer service for assistance.

Frequently Asked Questions

Here are some frequently asked questions about the Fire TV stick:

- How do I set up my Fire TV stick?
- To set up your Fire TV stick, simply plug it into an HDMI port on your TV, connect it to a power source, and follow the on-screen instructions to connect to your WiFi network and sign in to your Amazon account.

- How do I connect my Fire TV stick to a new WiFi network?
- To connect your Fire TV stick to a new WiFi network, go to the settings menu, select "Network," and then select "WiFi." Choose the new network and enter the password to connect.

- How do I update my Fire TV stick?
- To update your Fire TV stick, go to the settings menu, select "Device," and then select "About." Check for any available software updates and install them.

- Can I use my Fire TV stick on multiple TVs?
- Yes, you can use your Fire TV stick on multiple TVs by unplugging it from one TV and plugging it into another.

- Can I use my Fire TV stick outside of the US?
- The Fire TV stick is primarily intended for use in the US, and its features and content may vary depending on your location. However, you may be able to use it in other countries with a VPN service.

- How do I troubleshoot my Fire TV stick?
- If you're experiencing issues with your Fire TV stick, try restarting the device and your router, checking for software updates, and checking the Amazon Fire TV support page for troubleshooting tips and solutions.

- How can I improve the performance of my Fire TV stick?
- To improve the performance of your Fire TV stick, make sure it's updated to the latest software version, close any open apps that you're not using, and restart your router.

- How do I remove apps from my Fire TV stick?
- To remove apps from your Fire TV stick, go to the settings menu, select "Applications," and then select "Manage Installed Applications." Choose the app you want to remove and select "Uninstall."

- How do I connect my Fire TV stick to Alexa?
- To connect your Fire TV stick to Alexa, make sure your Fire TV stick and Alexa-enabled device are connected to the same WiFi network. Then, go to the Alexa app, and link your Fire TV stick under the settings.

- Can I mirror my smartphone screen to my Fire TV stick?
- Yes, you can mirror your smartphone screen to your Fire TV stick by using the screen mirroring feature in the settings menu or by using a third-party app such as "AllCast" or "AirScreen".

- Can I use a VPN with my Fire TV stick?
- Yes, you can use a VPN with your Fire TV stick by installing a VPN app on the device or by configuring a VPN on your router. This will allow you to access geo-restricted content and protect your online privacy.

- Can I use my Fire TV stick without a TV?

- Yes, you can use your Fire TV stick without a TV by connecting it to a compatible monitor or projector. You'll need to have an HDMI port on the monitor or projector and a power source.

- Can I use my Fire TV stick to watch live TV?
- Yes, you can use your Fire TV stick to watch live TV with the help of streaming services such as Sling TV, YouTube TV or Hulu TV. You can also use a TV tuner or a streaming device such as HDHomeRun to watch live TV on your Fire TV stick.

- Can I use my Fire TV stick to play games?
- Yes, you can use your Fire TV stick to play games by downloading games from the Amazon Appstore or by streaming games using a service like PlayStation Now or xCloud

- Can I use my Fire TV stick with a universal remote?
- Yes, you can use your Fire TV stick with a universal remote by programming the universal remote to control the Fire TV stick's functions. This can be done by using the manufacturer's instructions or by using a remote-control app on your smartphone.

- How can I improve the picture quality on my Fire TV stick?
- To improve the picture quality on your Fire TV stick, make sure your TV is set to the correct resolution and that the HDMI cable is securely connected. Additionally, you can adjust the display settings on your Fire TV stick by going to the settings menu and selecting "Display & Sounds" or "Display & Sound" depending on the model. From there, you can adjust the resolution, overscan, and other settings to optimize the picture quality.

- Can I use my Fire TV stick with a surround sound system?
- Yes, you can use your Fire TV stick with a surround soun system by connecting it to the system via HDMI or an optica audio cable. You can also connect wirelessly using Bluetooth c Wi-Fi.

- How do I turn off the Fire TV stick?
- To turn off the Fire TV stick, you can go to the settings menu select "Device" and then "Power" and choose the "sleep" optio Additionally, you can also unplug the power cord from the wa outlet to turn it off.

- Can I use my Fire TV stick to control other devices in m home?
- Yes, you can use your Fire TV stick to control other devices i your home by using the Alexa voice control feature. This allow you to use your Fire TV stick to control compatible smart hom devices, such as lights, thermostats, and more.

- Can I use my Fire TV stick to access other streaming services?
- Yes, you can use your Fire TV stick to access other streamin services such as Netflix, Hulu, Disney+, and many more b downloading the respective apps from the Amazon Appstore.

Contacting Amazon Customer Service

If you are experiencing issues with your Amazon Fire TV Stick 4K Max that cannot be resolved through troubleshooting, you can contact Amazon customer service for assistance. Here's a detailed look at how to contact Amazon customer service for help with your Fire TV Stick 4K Max:

- Go to the Amazon website: Open your web browser and go to the Amazon website.

- Sign in to your account: Sign in to your Amazon account using your email address and password.

- Find the "Help" button: Once you're signed in, look for the "Help" button, usually located at the top-right corner of the screen. Click on it.

- Select "Contact Us": Once you're in the Help section, you'll see a "Contact Us" button, click on it.

- Choose the appropriate category: Select the category that best describes your issue, such as "Fire TV Stick" or "Alexa"

- Choose the appropriate subcategory: Select the subcategory that best describes your issue, such as "Troubleshoot" or "Set up"

- Choose a way to contact: You will be provided with different ways to contact Amazon customer service, such as phone, email, or chat.

- Follow the instructions: Follow the instructions provided to you, such as providing your device's serial number, or explaining the issue in detail.

It's worth mentioning that, Amazon customer service is available 24/7, and that depending on the issue, you may be required to provide additional information to help them assist you better.

By contacting Amazon customer service, you can get expert assistance with resolving any issues you may be experiencing with your Fire TV Stick 4K Max. Customer service can guide you through troubleshooting steps and provide further help if needed.

Chapter 15: Conclusion

In conclusion, the Amazon Fire TV Stick 4K Max is a powerful and versatile streaming device that allows users to access a wide variety of streaming services, including Netflix, Prime Video, Disney+, and more. With its built-in Alexa voice control, 4K Ultra HD streaming capabilities, and a user-friendly interface, the Fire TV Stick 4K Max is a great way to enjoy your favorite movies and TV shows in high definition. Additionally, it allows the user to customize the device by changing the background, setting parental controls, and manage subscriptions. If you encounter any issues with your Fire TV Stick 4K Max, troubleshooting steps and contacting Amazon customer service are available to help you resolve them. With its range of features, the Fire TV Stick 4K Max is a great option for anyone looking to enhance their streaming experience.

Glossary: Terms and Definitions related to Fire Tv Stick 4k Max

Fire TV Stick 4K Max: A streaming device developed by Amazon that allows users to access a variety of streaming services, such as Netflix, Prime Video, Disney+ and others. It also includes Alexa voice control, 4K Ultra HD streaming capabilities, and a user-friendly interface.

Alexa: A virtual assistant developed by Amazon that allows users to control their Fire TV Stick 4K Max using voice commands. Alexa can be used to play and pause content, change the volume, search for content, and more.

HDMI: High-Definition Multimedia Interface, a type of audio/video interface used to transfer high-definition video and audio from a device to a TV or monitor. The Fire TV Stick 4K Max connects to the TV via HDMI.

Wi-Fi: A wireless networking technology that allows devices to connect to the internet. The Fire TV Stick 4K Max requires a Wi-Fi connection to access streaming services.

Streaming: The process of delivering digital media, such as movies and TV shows, over the internet in real-time. Streaming services, such as Netflix and Prime Video, can be accessed on the Fire TV Stick 4K Max.

Parental controls: A feature on the Fire TV Stick 4K Max that allows users to restrict access to certain content and apps based on their rating or type. This feature can be used to ensure children are not exposed to inappropriate content.

Subscriptions: A feature on the Fire TV Stick 4K Max that allows users to manage their subscriptions to streaming services such as Netflix, Prime Video, Disney+ and others. This feature allows users to modify, cancel or discover new content.

Screen mirroring: A feature on the Fire TV Stick 4K Max that allows users to mirror the screen of their mobile device or computer onto their TV. This feature allows users to share their device's screen with others on the TV.

4K Ultra HD: A high-definition video format that offers a resolution of 3840x2160 pixels, four times greater than standard 1080p HD. The Fire TV Stick 4K Max supports 4K Ultra HD streaming capabilities.

Next steps for users to take after reading the guide

After reading this guide, users should have a clear understanding c how to set up, use, customize, and troubleshoot their Amazon Fire T Stick 4K Max. The next steps for users to take would be to:

- Set up their Fire TV Stick 4K Max by following the instruction provided in the guide.

- Connect their Fire TV Stick 4K Max to their TV and Wi-F network.

- Register their Fire TV Stick 4K Max by creating an Amazo account or logging in with an existing account.

- Explore the home screen, navigate through the different menu and start streaming content.

- Use Alexa voice control to search for content and control the Fire TV Stick 4n K Max.

- Customize their Fire TV Stick 4K Max by changing th background, setting parental controls and managin subscriptions.

- Use screen mirroring to share their device's screen with othe on the TV.

- Keep the guide handy in case they encounter any issues ar refer back to it for troubleshooting steps.

- Contact Amazon customer service if they encounter any issues that cannot be resolved through troubleshooting.

By following these next steps, users will be able to fully set up and enjoy their Amazon Fire TV Stick 4K Max and get the most out of their streaming experience.

Made in the USA
Las Vegas, NV
29 December 2023